Urban Health and Wellbeing

Systems Approaches

Series Editor

Yongguan Zhu, Chinese Academy of Sciences, Institute of Urban Environment,
Xiamen, Fujian, China

The Urban Health and Wellbeing: Systems Approaches series is based on a 10-year global interdisciplinary research program developed by International Council for Science (ICSU), and sponsored by the InterAcademy Partnership (IAP) and the United Nations University (UNU). It addresses up-to-date urban health issues from around the world and provides an appealing integrated urban development approach from a systems perspective. This series aims to propose a new conceptual framework for considering the multi-factorial and cross sectorial nature of both determinants and drivers of health and wellbeing in urban populations and takes a systems approach for improving the understanding of the interconnected nature of health in cities. The systems approach includes an engagement with urban communities in the process of creating and transferring knowledge. Further, it aims at generating knowledge and providing the evidence that is relevant to people and policy-makers for improving integrated decision making and governance for the health and wellbeing of urban dwellers. The methods applied, come from various epistemological domains in order to improve understanding of how the composition and functioning of urban environments impacts physical, mental and social heath and how inequalities can be reduced to improve the overall quality of urban life.

The systems approach is applied to science and society and defined by a deep investigation into disciplinary knowledge domains relevant for urban health and wellbeing, as well as an inter- and transdisciplinary dialogue and shared understanding of the issues between scientific communities, policy makers and societal stakeholders more broadly. It involves one or more of the following elements: 1) the development of new conceptual models that incorporate dynamic relations among variables which define urban health and wellbeing; 2) the use of systems tools, stimulation models and collaborative modelling methods; 3) the integration of various sources and types of data including spatial, visual, quantitative and qualitative data.

Like the first book, the coming books will all address the topic of urban health and wellbeing, specifically by taking a systems approach. The topics range across all urban sectors and can, for example, cover the following areas:

(1) transportation, urban planning and housing, urban water, energy and food, communication, resources and energy, urban food systems, public service provision, etc.
(2) the related health disorders in physical, social and mental health
(3) the methods and models used and the type of science applied to understand the complexity of urban health and wellbeing.

More information about this series at http://www.springer.com/series/15601

Franz W. Gatzweiler

Editor

Urban Health and Wellbeing Programme

Policy Briefs: Volume 2

Editor
Franz W. Gatzweiler
Institute of Urban Environment
Chinese Academy of Sciences
Xiamen, China

ISSN 2510-3490   ISSN 2510-3504   (electronic)
Urban Health and Wellbeing
ISBN 978-981-33-6038-9   ISBN 978-981-33-6036-5   (eBook)
https://doi.org/10.1007/978-981-33-6036-5

Jointly published with Zhejiang University Press
The print edition is not for sale in China (Mainland). Customers from China (Mainland) please order the print book from: Zhejiang University Press.

This Springer imprint is published by the registered company Springer Nature Singapore Pte Ltd.
The registered company address is: 152 Beach Road, #21-01/04 Gateway East, Singapore 189721, Singapore

Foreword

We live on an urban planet of increasing interconnectivity and complexity. Cities are the new melting pots of global development. Over half of the world's population lives in cities, and this number is increasing by about 2% annually. More than two billion urban dwellers are expected to be added over the next three decades, a significant proportion of whom will be living in informal or slum settlements. Urban areas are extremely complex environments in which environmental, social, cultural and economic factors influence people's health and wellbeing and have an impact on planetary health. To cope with this complexity, understand it and find sustainable solutions, the Urban Health and Wellbeing (UHWB) programme has developed the systems approach and promotes systems thinking for improving the health and wellbeing in and of cities worldwide.

The Urban Health and Wellbeing Programme (UHWB) is a global science programme and interdisciplinary body of the International Science Council (ISC, previously ICSU), supported by the International Society for Urban Health (ISUH), the InterAcademy Partnership (IAP). The overarching vision for UHWB are people to develop aspired levels of wellbeing by living in healthy cities.

This book is a collection of policy briefs produced from research presented at the 16th Conference on Urban Health in Xiamen, China, November 4–8, 2019 under the theme "People Oriented Urbanisation: Transforming Cities for Health and Well-Being", co-organized by the Urban Health and Wellbeing (UHWB) programme of the International Science Council (ISC).

The UHWB programme takes an interdisciplinary, cross-sectoral and systemic view on issues of health and wellbeing in cities which include the urban economy and finance systems, education, employment, mobility and transport, food, energy and water resources, access to public services, urban planning, public spaces and urban green, as well as social inclusion. Contributions to this book have been made by scientists from multidisciplinary research fields.

The policy briefs in this volume present the background and context of an urban health issue, research findings and recommendations for policy/decision-makers and action-takers. In some cases they inform about relevant events and developments from the science community or important opinion pieces which address health emergencies, like the current COVID-19 pandemic.

This book is intended for citizens and political decision-makers who are interested in systems perspectives on urban health and wellbeing, examples of how to deal with the increasing complexity of cities and the accompanying environmental and social impacts of increasing urbanization. Furthermore, it hopes to inspire decision-makers to facilitate finding solutions, in order to reach the goal of advancing global urban health and wellbeing.

Xiamen, China   Franz W. Gatzweiler
August 2020   Yu Liu

Contents

COVID-19, Cities and Health: A View from New York

Jo Ivey Boufford and Anthony Shih

1 Key Messages

1. Four characteristics of cities: density, diversity, complexity and disparities are helpful for analysis of problems.
2. The COVID-19 pandemic has more publicly revealed existing vulnerabilities in urban life when facing such an emergency.
3. A systems approach to governance across all sectors is critical to identifying problems and developing interventions to address them.
4. A commitment to governance for health by political leadership can clarify options for action in complex systems.

Population density is one of the defining characteristics of cities and is also one of the dominant challenges in addressing COVID-19 and likely future pandemics. In normal times, urban density allows for cost-effective service delivery, acts as a magnet for jobs and the generation of wealth, and nurtures innovation. Coupled with diversity—another core characteristic of cities—urban density also yields the cultural variety in urban neighbourhoods, fosters global connections and makes cities like New York attractive and exciting. However, in the midst of a pandemic, the very thing we love about cities becomes their apparent weakness: our urban density allows for more rapid spread of infectious disease and makes policy interventions such as social distancing much more difficult to follow—even with the best of intentions. Our

J. I. Boufford (✉)
Global Health, New York University School
of Global Health, New York, NY, USA
e-mail: jboufford@isuh.org

A. Shih
United Hospital Fund (UHF), New York, NY, USA
e-mail: ashih@uhfnyc.org

© Zhejiang University Press 2021
F. W. Gatzweiler, *Urban Health and Wellbeing Programme*,
Urban Health and Wellbeing,
https://doi.org/10.1007/978-981-33-6036-5_1

1

cultural diversity can also make the need for such interventions more challenging to communicate and manage.

This has all led some commentators to speculate about the death of cities, including the potential for a mass exodus once the pandemic is over. While this is not an economic option for most, it also ignores the reality that urbanization is a major global trend that will likely not be stopped by this or another pandemic. Four out of five Americans live in urban areas; across the world, over half of the population does, and the number is growing. In the midst of a pandemic, it's easy to focus on the risks of living in cities. What's not as easy—but just as critical—is ensuring that we learn from this experience and consider how best to address these challenges in the future and how to prevent or greatly reduce the negative health and socio-economic effects of pandemics like COVID-19.

Two other characteristics of cities have been thrust in the spotlight by COVID-19. One is complexity—the interdependence of the multiple systems that make cities run well and the challenge of governing across sectors for better health—a systems approach to problem-solving and decision-making. For example, in transportation policy, mass transit can act as a hub for the spread of coronavirus but also has a positive effect on air quality and related reductions in rates of asthma. Housing policy can either advance or impede safe and affordable dwellings. The other key urban characteristic is the existence of disparities within cities. The wide discrepancies in mortality from COVID-19 in New York City can be mapped to geographic communities that have long suffered from structural inequalities in housing, education, access to parks, recreation and healthcare, food security and socio-economic opportunities.

One positive outcome emerging from this pandemic is a broader appreciation for and investment in the public health infrastructure that monitors and reports outbreaks, provides critical public information and tests and tracks infection and immunity in the community. Our hospitals and healthcare workers have been amazing in their responses and we must be sure this resource stays strong. But as we think about the configuration of our healthcare system going forward, we in the United States need a healthcare financing system which supports hospitals to do what they do best— manage emergencies and diagnose and treat complex diseases—while assuring additional investment in primary care, telemedicine and more support for critical community-based organizations that have the trust of communities and can provide crucial social support to families.

While these policy changes within the health sector would be wonderful and welcome, COVID-19 is demonstrating to the public what urban health researchers, practitioners and policymakers have known for decades—the importance of a systems approach in understanding and acting on the determinants of health. We hope for a newfound appreciation of how all the sectors that form the core of city life affect the health and wellbeing of New Yorkers and how a systems perspective to governing for health reveals that: housing policy is health policy, transportation policy is health policy, education policy is health policy and political leadership for health improvement can make cities engines to improve national and global health. There are examples of cities around the world that have taken on and successfully met these challenges.

As a global city, New York can set a bold example. We can go from being the epicentre of a pandemic to an international model of a city whose systems and leadership act to ensure the health of all its residents.

Current and Future Human Exposure to High Atmospheric Temperatures in the Algarve, Portugal: Impacts and Policy Recommendations

André Oliveira, Filipe Duarte Santos, and Luís Dias

1 Key Messages

1. There is currently evidence of increased morbidity and mortality associated with high atmospheric temperatures. Risk factors such as age, ethnicity and behavioural factors, among others, tend to aggravate these impacts. Urban areas generally present an increased risk, both outdoor (mainly due to the urban heat island effect) and indoor, due to low housing standards (such as poor insulation).
2. High-temperature episodes are currently the most relevant phenomenon to human health in a context of climate change, considering the projected trend of temperature rise, estimated to increasingly affect the south of the European continent. Additionally, increasing population exposure is expected, particularly during the summer period.
3. This trend is critical for the Algarve region, considering its hot summer, with a current average monthly maximum temperature of 28 °C in July and August, projected to worsen progressively under climate change scenarios over the century, particularly in the most severe scenario (i.e. RCP8.5) and the 2071–2100 period.

A. Oliveira (✉) · L. Dias
Climate Change Impacts, Adaptation and Modelling Research Group (CCIAM), cE3c Research Centre, Faculty of Sciences, University of Lisbon, Lisbon, Portugal
e-mail: mafoliveira@fc.ul.pt

L. Dias
e-mail: lfdias@fc.ul.pt

F. D. Santos
Climate Change and Sustainable Development Policies, University of Lisbon and New University of Lisbon, Lisbon, Portugal
e-mail: fdsantos@fc.ul.pt

© Zhejiang University Press 2021
F. W. Gatzweiler, *Urban Health and Wellbeing Programme*,
Urban Health and Wellbeing,
https://doi.org/10.1007/978-981-33-6036-5_2

4. Policymakers should prioritize the implementation of evidence-based adaptation measures, to efficiently increase the resilience of human population to high-temperature episodes. Furthermore, measures should include a focus on critical populations segments, such as the poor and the elderly.

2   Human Health and High Atmospheric Temperatures

Epidemiological evidence of increased mortality and morbidity during heatwaves or other episodes of high atmospheric temperature has been established, as well as the contribution of several risk factors, such as ageing, behaviour, previous health conditions (e.g. cardiovascular and respiratory disease) and socio-economic factors (Basu and Samet 2002; Loughnan et al. 2012; Sanderson et al. 2017). Episodes of high atmospheric temperature are expected to continue increasing, and the population exposed to heat extremes has experienced a major increase since 1990 (Smith et al. 2014; Watts et al. 2018).

Due to the urban heat island effect, urban inhabitants are exposed to higher temperatures than rural populations. Future urban temperatures can rise due to the warming climate and the expansion of urban centres. Many climate models do not simulate urban climates, so current projections of future heat-related mortality within cities are likely underestimated. Other pitfalls contribute to uncertainty in projections of future heat-related mortality, such as misuse of climate models and emissions scenarios; calibration methods; and the way population issues like adaptation to warmer temperatures, future population and demographic changes (particularly ageing) are included (EASAC 2019; Sanderson et al. 2017).

In Portugal, current heat mortality risk is evaluated, for routine surveillance and early warning purposes, using the ICARO surveillance system. This is based on excess mortality estimates under a high-temperature forecast for the following three days, calculated for the Portuguese mainland and its five NUTII regional divisions (Nogueira and Paixão 2008; Nogueira et al. 2010).

Few studies have included risk projections for Portugal. A recent study included the city of Lisbon (point representation, not regionalized) and classified it among those in the top 25% maximum temperature during a heatwave (Guerreiro et al. 2018), suggesting a need for further research on this topic.

This study aims to contribute to a better preparedness of the Algarve region (the southernmost Portuguese mainland region) in a context of climate change; it was executed within the framework of the Algarve Intermunicipal Climate Change Adaptation Plan—PIAAC-AMAL (Dias et al. 2019; Oliveira et al. 2019).

3 Current and Projected Human Health Impact of Hot Days in the Algarve

We calculated projections of future mortality risk for all municipalities of the Algarve region using spatial statistical methods, historical measurements and future temperature projections under RCP4.5 and 8.5 (IPCC 2013), to model non-accidental mortality risk (% increase of deaths) as a function of the number of hot days (maximum temperature above 30 °C) and sociodemographic variables, namely % elderly municipal population (residents above 65 years old), % municipal *per capita* purchasing power and % municipal urban area.

Currently, average annual maximum temperature varies between 15.6 and 23.4 °C, influenced by mountainous areas along central Algarve and the Atlantic Ocean in western Algarve (*Barlavento*) whereas the highest values occur along the Guadiana river in eastern Algarve (*Sotavento*).

Under climate change scenarios, the Sotavento area may come to experience an increase in average maximum temperature close to 4 °C by the end of the century. Tropical nights (days with minimum temperature exceeding 20 °C) range currently between 9 and 16 days but can increase up to 66 more days under RCP8.5 scenario by the end of the century in coastal Sotavento, comparatively to the current situation.

Concerning the number of hot days, considerable increases are projected for the end of the century throughout the region. Northernmost Sotavento, near the Guadiana river, experiences currently around 70 days a year, but this count can rise to approximately 50 more days (Fig. 1).

A higher incidence and longer average duration of heatwaves is projected, especially in inland areas. In the last 30 years, 2–13 heatwaves were registered, with an average period between 6 and 9 days (Barlavento and Sotavento, respectively). Up to 200 heatwave events may occur in 2071–2100 (RCP8.5), with an average period longer than 4 days, compared to the current period.

According to our current mortality model, Algarve municipalities had, on average, 2% of non-accidental heat-related mortality, during the reference period adopted (1991–2003).

In line with projected increases in hot day counts, mortality projections (Fig. 2) signal a progressive rise in mortality risk over the considered periods, under scenarios RCP4.5 and RCP8.5, with the risk gradually increasing from east to western Algarve. The worst situation (near 8% increase) is expected by the end of the century under RCP8.5, at the north-eastern tip of the territory.

4 Policy Recommendations

To transpose our findings into policy recommendations, projected impacts and adaptation measures were evaluated in the light of existing evidence, and comprehensive stakeholder consultations were performed. Resulting from this process, targeted

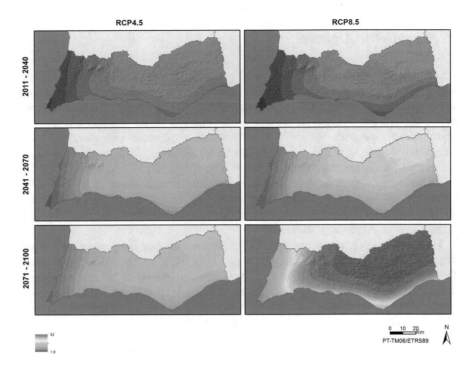

Fig. 1 Annual count of days 30 °C (anomaly in nº of days)

policy recommendations (designed as adaptation measures) were proposed for the Algarve region. For a comprehensive insight on proposed adaptation measures, the Algarve Intermunicipal Climate Change Adaptation Plan (Dias et al. 2019) should be consulted.

The urban environment is a system, and therefore all policy recommendations presented here were designed to be implemented as such, in the sense that they are either complementary among themselves, or at the very least minimize or avoid negative impacts to other measures or the system as a whole.

In many situations, the first step to protect a population is to guarantee that persons are adequately informed to protect themselves. Therefore, it is imperative to increase the level of knowledge in the general population (particularly among the most vulnerable groups), visitors (e.g. tourists) and public and private institutions, regarding the adoption of good practices when dealing with episodes of high atmospheric temperature. This can be achieved by promoting informative initiatives, such as public sessions and making use of social media and networks.

Beyond purely informative initiatives, a critical recommendation is to strengthen and empower healthcare delivery systems to be able to deal with the progressive increase of high-temperature episodes. Among other initiatives, a comprehensive spatial database of persons included in vulnerable groups living alone (e.g. in urban

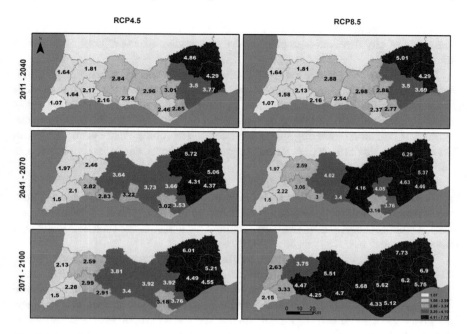

Fig. 2 Mortality increase (%) projected according to counts of days ($T > 30$ °C) under climate change

areas) or in isolation (e.g. in rural mountain areas) can be implemented. These persons can then be regularly monitored by health professionals, especially before high-temperature episodes, so that emergency relocation to climatized spaces (e.g. shopping malls) can be considered promptly.

Preparing the built environment is also essential. This encompasses acting both at the level of urban design (e.g. planning for urban wind corridors) and to improve housing standards, with a focus on socio-economically deprived citizens.

Going beyond conventional infrastructure interventions, another way to foster the resilience of city dwellers is to develop bioclimatic studies of public space (current and future) and implement bioclimatic architecture solutions in new buildings, façades and existing roofs (e.g. green roofs).

Another less conventional intervention is to control air temperature using micro-sprinklers. Although these devices have a limited range, they can be beneficial. These devices can be deployed as corridors in unprotected urban spaces (e.g. to assist traversing large squares without shadows or green space), or in a more localized manner, to lower air temperature in outdoor spaces (e.g. open terraces of bars, cafeterias or smoking areas). Care should be taken with the water supply and maintenance of these devices, to avoid outbreaks of aerosol-carried infections (e.g. Legionnaire's disease).

A critical component of urban life is mobility. Therefore, a form of improving the resilience of city dwellers is to place active and passive means of cooling in public transport.

Increasing urban green area (using autochthonous species whenever possible) improves thermal comfort and can be very beneficial to reduce the impacts of other threats which are themselves intensified by high temperatures, such as urban atmospheric pollution. Structurally, green areas can be deployed as a network of urban ecological corridors, fostering sports practice such as running and cycling, and with other positive side effects (e.g. stimulating associated commercial activities such as cafeterias and bike renting shops).

It is essential to provide shading areas. This can be achieved using vegetation or artificial materials when the former is not possible, such as in historical, densely built street networks. For instance, fabric street covers can be deployed during summertime, a solution already implemented in some Portuguese cities such as Águeda and Loulé, the latter in the Algarve.

The introduction of artificial structures to promote the presence of water in public space is also recommended to decrease air temperature, albeit within a limited effective range. To improve cooling efficiency and avoid the establishment of vector-borne diseases, precautions should be taken to guarantee water movement. For instance, the presence of water can be implemented in the form of an urban stream network, if possible, by mimicking natural hydrological systems.

References

Basu R, Samet J (2002) Relation between elevated ambient temperature and mortality: a review of the epidemiologic evidence. Epidemiol Rev 24:190–202

Dias LF, Aparício B, Veiga-Pires C et al (2019) Plano intermunicipal de adaptação às alterações climáticas do Algarve, CI-AMAL (PIAAC-AMAL). Faro, Portugal

EASAC (2019) The imperative of climate action to protect human health in Europe

Guerreiro SB, Dawson RJ, Kilsby C et al (2018) Future heat-waves, droughts and floods in 571 European cities. Environ Res Lett 13:34009

IPCC (2013) Climate change 2013: the physical science basis. Contribution of working group I to the fifth assessment report of the intergovernmental panel on climate change. Cambridge, United Kingdom and New York, NY, USA

Loughnan M, Nicholls N, Tapper NJ (2012) Mapping heat health risks in urban areas. Int J Popul Res 2020:1–12

Nogueira P, Paixão E (2008) Models for mortality associated with heatwaves: update of the Portuguese heat health warning system. Int J Climatol 28:545–562

Nogueira PJ, Machado A, Rodrigues E et al (2010) The new automated daily mortality surveillance system in Portugal. Eurosurveillance 15:pii=19529

Oliveira A, Dias LF, Aparício B (2019) Relatório Setor: Saúde Humana: Vulnerabilidades Atuais e Futuras. Faro, Portugal

Sanderson M, Arbuthnott K, Kovats S et al (2017) The use of climate information to estimate future mortality from high ambient temperature: a systematic literature review. PLoS ONE 12:e0180369

Smith KR, Woodward A, Campbell-Lendrum D et al (2014) Human health: impacts, adaptation, and co-benefits. In: Climate change 2014: impacts, adaptation, and vulnerability. Part A: global and sectoral aspects. Contribution of working group II to the fifth assessment report of the intergovernmental panel of climate change. Cambridge University Press, Cambridge, United Kingdom and New York, NY, USA, pp 709–754

Watts N, Amann M, Arnell N et al (2018) The 2018 report of the Lancet Countdown on health and climate change: shaping the health of nations for centuries to come. Lancet (London, England) 392:2479–2514

Neuroscience-Based Urban Design for Mentally Healthy Cities

Agnieszka Olszewska-Guizzo

1 Key Messages

1. Mental health is an extremely important topic in our urbanized world. Due to the wide inaccessibility of mental healthcare (isolation, high therapy costs, stigma), new non-pharmacological, self-care approaches are needed.
2. Connection with nature proves to contribute significantly to the improvement of mental health of urban dwellers. More specifically, it is the quality and accessibility of urban green spaces—the mere quantity and area of green space in the city may not be enough.
3. There are certain types and characteristics of urban green spaces that can elicit beneficial brainwave patterns in the human brain. Anyone can experience this through nothing more than a passive exposure to them.
4. The grouping of these beneficial characteristics of green spaces was coined contemplative, and the brain response to them can be associated with stress reduction, emotion regulation and even helping with depression.
5. The most contemplative views within the city should be identified and protected; landscape architects and urban planners should include contemplative landscape features within their designs to create more mentally healthy cities.

2 Background

Urbanization and the recent COVID-19 pandemic highlight the pressing issue of mental health in the urbanized world. It is vital to recognize urban green spaces as a

A. Olszewska-Guizzo (✉)
National University of Singapore, Singapore, Singapore
e-mail: a.o.guizzo@neurolandscape.org

© Zhejiang University Press 2021
F. W. Gatzweiler, *Urban Health and Wellbeing Programme*,
Urban Health and Wellbeing,
https://doi.org/10.1007/978-981-33-6036-5_3

13

medium to improve mental health and wellbeing for city residents to create healthy cities through mentally healthy living environments.

A growing body of evidence from observational and experimental studies shows the associations between exposure to urban green spaces and mental health outcomes. Many urban greening practices have been taking place around the world, but the research shows that quantity of green is not as important for mental health delivery as the quality of the green space. The presented line of research attempts to uncover the specific features of green spaces that are the most beneficial for human mental health, utilizing the objective physiological markers of mental health, such as the assessment of brain activity.

The recent EEG experiment was conducted in three different urban spaces of the high-density city-state of Singapore: (1) urban park, (2) neighbourhood roof garden, and (3) busy urban area with scenes including pedestrian plaza, street and transport hub (Fig. 1). (1) and (2) represent green spaces with different views and designs and (3) served as the control.[1]

3 Findings

The preliminary evidence has shown that the passive exposure to certain landscape compositions (termed contemplative landscapes) triggered positive emotions and motivation during passive exposure. This finding is supported by objective EEG measurements on brain activity, through the exhibition of frontal alpha asymmetry (FAA) (Fig. 2a). Even within the individual sites, there may be a heterogeneous response to different views and some scenes may induce more or less FAA response than others. Furthermore, there are preliminary indications that females show higher responsiveness to change in green space quality (Fig. 2b).

The contemplative landscape features deemed the most beneficial for mental health and wellbeing consist of aggregation of multiple aspects within a view. The main contemplative features of these views include: far-away view, which permits seeing the depth of the landscape; undulating landform and diversified skyline, which brings the focus towards the sky; shaded observation points and design which permits perception of the changes of daily and seasonal cycles; vegetation seemingly rich in species and self-sown, with a tamed degree of wilderness; balance and harmony in design and composition with a lack of disturbing elements, balanced scales of physical objects and view openings; explicit presence of archetypal elements, such as a water mirror, waterfall, boulder, clearing, single old tree, among others; lastly, the strong character of peace and silence of the scene, with a sense of solitude and reorientation of the urban context with the design inviting rest and relaxation.

---

[1] Olszewska-Guizzo et al. (2020).

Fig. 1 Experimental sites around Singapore with three scenes within each of them. Site 1 and 2 are green spaces (contemplative landscape score in 1–6 point scale provided); Site 3 is a typical busy urban environment with street traffic, urban plaza and crowds, which served as the control

4 Conclusions and Recommendations

This study investigated the effects of passive exposure to different urban scenes, without engaging participants in any additional tasks or activities. It shows that the brain response was different depending on the quality of the green scenes. To achieve the most beneficial mental health outcomes from passive exposure, the exposure to contemplative landscape features is recommended. These can be identified and/or introduced in our cities with the existing Contemplative Landscape Model (CLM), which has been validated in the urban landscape discipline. Moreover, the findings suggest that contemplative design, being a background to everyday urban living, can benefit the overall mental health and wellbeing of the urban population. On the other hand, mindless city greening, often used to meet the spatial quota, may not be effective in reaching that goal.

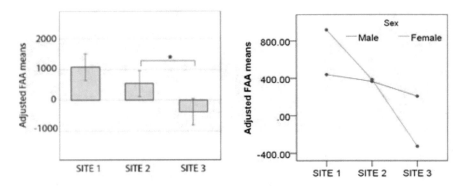

Fig. 2 Frontal alpha asymmetry across participants passively watching. **a** three different sites: urban park with scenes of high contemplative value (Site 1), neighbourhood green roof park (Site 2) and urban busy street area (Site 3). **b** presented separately for males and females

Landscape designers and urban planners should incorporate the insights of evidence-based design, such as contemplative landscape design strategies, in their vision of a healthy future city. Efforts should be put into translation of the empirical knowledge to design practice through publications and design workshops. Neuroscience-based design can provide novel information on the effects of the features of urban spaces on mental health and wellbeing to inform future interventions and prevention programmes.

Provision of equitable access to high-quality urban green space, especially for persons with disabilities, which can also be described as that clinical mentally ill population or non-clinical population with high stress exposure, is to be recognized as a determinant and a trigger for better mental health, aligned with the Sustainable Development Goals 11.7 and 11.3, and through its direct association with social and environmental justice—SDG 15.9.

Reference

Olszewska-Guizzo A, Sia A, Fogel A, Ho R (2020) Can exposure to certain urban green spaces trigger frontal alpha asymmetry in the brain?—preliminary findings from a passive task EEG study. Int J Environ Res Public Health 17(2):394

The Role of Money for a Healthy Economy

Felix Fuders

1 Key Messages

1. All but one of the SDGs are directly linked to our current financial system which—being completely unnatural—can be seen as the most important but at the same time least recognized reason for market failure.
2. Only if we change the unnatural design of our money to a more natural one as proposed 100 years ago by Silvio Gesell, we will be able to create a healthier economy and reach the goals.
3. Such a reform of our financial system will also reduce the possibility of the occurrence of major financial crisis and, furthermore, could be an alternative way to address the imminent economic crisis that threatens world economy as a result of the COVID-19 quarantine measures.

2 Background: The World Is on a Collision Course

If we take a look at any newspaper in any country, two major problems frequently are addressed: inequality and the increasing destruction of the natural environment, that is unsustainability in the *stricto* sensu. In 2015, more than 190 world leaders recognized that the world is on a "collision course" (Max-Neef 2010) and committed to 17 Sustainable Development Goals (SDGs). All but one goal are either linked to the unsustainability of our current lifestyle (goals 9, 11–15) or to inequality (goals 1–8, 10, 16). Many conferences and high-level meetings have been held since then

F. Fuders (✉)
Right Livelihood College—Campus Austral and Economic Policy Chapter of the
Transdisciplinary Research Center for Socio Ecological Strategies for Forest Conservation
(TESES), Universidad Austral de Chile, Economics Institute, Valdivia, Chile
e-mail: felix.fuders@uach.cl

© Zhejiang University Press 2021
F. W. Gatzweiler, *Urban Health and Wellbeing Programme*,
Urban Health and Wellbeing,
https://doi.org/10.1007/978-981-33-6036-5_4

and one of the major topics is how to finance these goals. However, it is not only more money that is required but a different kind of money. In fact, both the overexploitation of the natural environment and the income inequality are directly connected to our unnatural financial system, and especially a misunderstanding of what money is and what it should be.

3 Insights and Findings

3.1 Money Is Like the Blood of the Economy

Unlike real goods, money is easily storable and does, therefore, not easily circulate in the economy. However, this is what it is supposed to do. Money can be described as the blood of our economy. Blood needs to circulate, otherwise the body gets ill. Similar to the blood circulation in the human body also the economy gets ill if money does not circulate well. Money as a calculation unit is supposed to be a medium to facilitate the exchange of goods and services. But, because people tend to preserve what Keynes (1936: 165 ff., 194 ff.) called the *"preference for liquidity"* we like to save up money, the more the better. With real goods, hoarding would only be possible in a very restrictive way, since real goods perish. Any excessive hoarding would, in time, result in the loss of the hoarded goods. But our money, as it is designed today, makes it possible to hoard any surplus almost without restriction. The unnatural design of our current financial system makes possible the hoarding of values produced, which provides a strong incentive to produce more than is actually needed.

3.2 Our Monetary System Obligates to Grow

The design of our money does not only give a powerful incentive to produce more than is actually needed; it even **obligates to do so** and here, again, the driver is the unnatural storability of money (for similar argumentation see Kennedy 2011; Creutz 2018). Money stored under the pillow does not circulate and, therefore, cannot serve the economy. This puts the owner of money in a monopolistic-like position to either choke the economy or to blackmail the person in need of money and *"press"* interest (Gesell 1949: 205, 344). To put it in the words of Keynes (1936: 167), interest is a *"reward for parting with liquidity"*, the incentive to lend the money out (or to bring it to the bank which then lends it out for us). To come back to the above-used metaphor of money as the "blood of the economy", interest can be seen as the "drug" that makes money circulate in the economy. And as any drug that is applied for a prolonged period of time, also the money interest rate brings along heavy side effects, especially the obligation to grow and inequality.

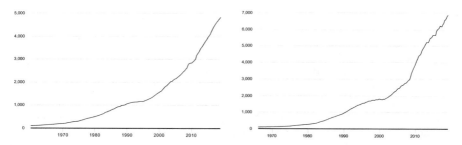

Fig. 1 Left: Estimated money supply M3 U.S. (billions USD) (*Source* Own work, based on data provided by OECD). Right: Federal debt U.S. (billions USD) (*Source* Own work, based on data provided by the US department of treasury)

Interest makes deposits grow and, since there is no interest without debt, also the total debt of an economy has to grow in a likewise manner. This is a simple fact. To say it in the words of Frederick Soddy (1934: 25) "Money is a credit-debt relation from which none can effectually escape". And even worse: money supply and debt do not only grow linearly but following an exponentially function. This is because any amount deposited on an interest-bearing bank account will have doubled after some time. Here we understand why the money supply (defined as bank deposits plus cash) in any country grows exponentially and so does total debt as we can observe in Fig. 1 that depicts money supply and total debt for the United States. We will recognize the same exponential function of both money supply and debt in any country of the world if the observed time period is only long enough.

A steady increase of the total debt means that we are ever more indebted, and this is the reason why we have to work ever more to not lose status quo. We can corroborate this taking a look at companies' balance sheets and compare them with some 20 years ago. It is very likely that today we will find much more borrowed capital. But even if a business is not financed with borrowed capital it is not freed from the obligation to grow. This is because interest is the opportunity cost of any productive investment. Any business that does not yield a return at least as high as what the business owner could receive depositing the money on a bank account is economically inviable. Hence, interest is the rhythm to which the real (productive) economy has to dance, and not just a "fetish" (Hamilton 2003). Since economic growth (real GDP growth) means that this year we produce more than the year before, and since the first law of thermodynamics tells us that we cannot produce something out of nothing, a steady GDP growth rate in the long run **must** end up in an increased use of natural resources. Any so-called "green" politics that do not take into account our financial system can therefore be seen as a farce (Fuders and Max-Neef 2014).

3.3   *Our Monetary System Produces Income Inequality*

The money interest rate is also a powerful driver behind income inequality. As said above, money supply and debt grow in a likewise manner. This means that while on the one side we find ever more monetary units on bank accounts, on the other side ever more people are ever more indebted. In other words: the income inequality, too, grows exponentially. We can visualize this drawing money supply and debt on the same graph. In Fig. 2 using data from Germany we can see that monetary assets mirror total debt. This is no surprise since, as outlined above, there is no interest paying without debt. The gap between the upper and the lower point in the graph is the (exponentially growing) inequality. The same gap can be observed in any other country if the observed time period is only long enough. Accordingly, the Gini-index that measures inequality shows not only similarly high values for almost all OECD countries but also a significant incrementation of inequality over the last 30 years (Bárcena et al. 2018).

4   Policy Recommendation: Gesell's Solution

Economic schools should study and analyse the conventional economic theories in order to formulate a new model of a market economy that is not perverted by the need to grow and by a constantly increasing income inequality, that is a market economy with a different kind of money. We could probably learn a lot from the proposal offered by the German-Argentine economist Silvio Gesell (1949) in his work, "The natural economic order". Gesell designed a currency that being equipped with some sort of "carrying cost" (Keynes 1936: 357) cannot be hoarded eternally, and thus circulates without interest as reward for parting with liquidity being necessary. Money then loses its special position compared to real goods, and the money holder cannot "press" interest anymore (Gesell 1949: 205, 344). This interest-free currency, Gesell called it *"free money"*, would serve solely as a means to facilitate the interchange of goods and services and not to store wealth. Consequently, it would truly comply with the concept that conventional economic theory usually calls monetary neutrality, but which does not hold under today's financial system (see already Suhr 1989).

Probably the easiest way to practically employ Gesell's proposal would be through an effective negative Central Banks interest rates policy, i.e. a monetary policy where negative interest rates are not only charged for deposits of commercial banks at the Central Bank, but that also apply to cash. Different ways have been discussed to achieve this (Buiter 2005; Seltmann 2010; Agarwal and Kimball 2015; Assenmacher and Krogstrup 2018).

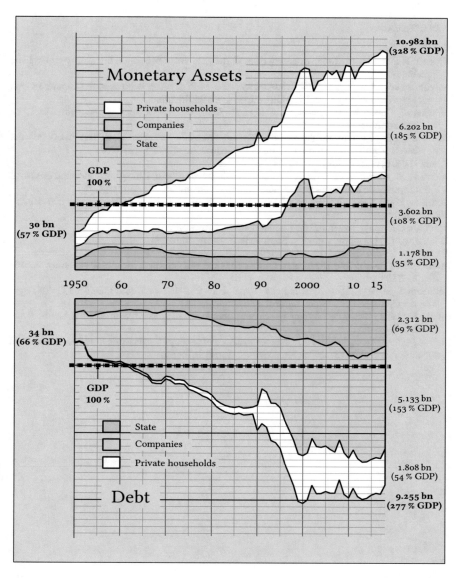

Fig. 2 Monetary assets and debt in Germany (*Source* Thomas Kubo/Helmut Creutz, based on data from Deutsche Bundesbank)

References

Agarwal R, Kimball M (2015) Breaking through the zero lower bound—IMF Workig Paper. IMF, Washington

Assenmacher K, Krogstrup S (2018) Monetary policy with negative interest rates: decoupling cash from electronic money—IMF Working Paper. IMF, Washington

Bárcena A, Cimoli M, García-Buchaca R et al (2018) The inefficiency of inequality. ECLAC, Santiago de Chile

Buiter W (2005) Overcoming the zero bound on nominal interest rates: Gesell's currency carry tax vs. Eisler's parallel virtual currency. Int Econ Econ Policy (IEEP) 2:89–200

Creutz H (2018) Das Geldsyndrom. Kubo, Münster

Fuders F, Max-Neef M (2014) Local money as solution to a capitalistic global financial crisis. In: From capitalistic to humanistic business. Palgrave Macmillan, Hampshire, NY

Gesell S (1949) Die natürliche Wirtschaftsordnung durch Freiland und Freigeld, 9th edn. Rudolf Zitzmann, Lauf

Hamilton C (2003) Growth fetish. Allen & Unwin, Crows Nest

Kennedy M (2011) Occupy money. Kamphausen Verlag, Bielefeld

Keynes JM (1936) General theory of employment, interest and money. Hartcourt, New York

Max-Neef M (2010) The world on a collision course and the need for a new economy. Ambio 39:200–210

Seltmann T (2010) Umlaufsicherung von Banknoten - Wie sich die Einführung der Liquiditätsgebühr bei Geldscheinen systemkonform realisieren lässt. Humane Wirtschaft 2:10–16

Soddy F (1934) The role of money—what it should be contrasted with what it has become. Routledge, London

Suhr D (1989) The capitalistic cost-benefit structure of money—an analysis of money's structural nonneutrality and its Effects on the economy. Springer, Berlin

Developing Health-Promoting Schools: An Initiative in Government Schools of Indore City, India

Alsa Bakhtawar

1 Key Messages

1. Young people constitute one of the precious resources of India. In a phase characterized by growth and development, youth are vulnerable to influence by factors that affect their health and safety. Addressing young people in their schools is an efficient way to reach them en masse.
2. Training of teachers as health ambassadors, promoting healthy school environments and facilitating multisectoral collaboration are key activities that can develop Health-Promoting Schools (HPS). These three activities will give opportunities to students to learn, adopt and practise hygiene and healthy lifestyles.
3. HPS have been successful in bringing multiple sectors together for their implementation which can be seen as a starting point for further collaborations in Indore whose goal is a "healthy, liveable Smart City for all".

2 Background

According to the 2011 Census, nearly 31% of India's population lives in urban areas (Government of India, n.d.). Looking at the trends of population growth, it is expected that urban areas will house 40% of India's population and will also contribute to 75% of India's GDP by 2030 (Sankhe et al. 2010). Preparing and adapting to this trend will require adequate development of the city's physical, social and economic infrastructure, and the development of Smart Cities model is a step in that direction. The Smart Cities Mission was introduced by the Government of India

A. Bakhtawar (✉)
Building Healthy Cities (BHC), New Delhi, India
e-mail: alsa_bakhtawar@in.jsi.com

© Zhejiang University Press 2021
F. W. Gatzweiler, *Urban Health and Wellbeing Programme*,
Urban Health and Wellbeing,
https://doi.org/10.1007/978-981-33-6036-5_5

23

in 2014 as an innovative step to drive economic growth and improve the quality of life of people by focusing on development and technology to create better outcomes for citizens. Health and education are included in the core infrastructural elements of the Smart City Mission. However, in most of the instances, these two elements have not received the desired attention by the city planners and administrators, and there have been limited cross-sectoral activities across these two departments.

Young people are a precious resource of India. In a phase characterized by growth and development, youth are vulnerable to influence by factors that affect their health and safety. As urban populations expand, so will the numbers of young people. As cities consider how best to reach them, interest in schools has increased. First advocated by the World Health Organisation, Health-Promoting Schools (HPS) is a concept of a school as a healthy setting for living, learning and working (World Health Organization, n.d.). In 1946, the Bhore Committee identified school health as an important health intervention (Kaur et al. 2015) to promote the maximum physical, social, emotional, mental and educational growth of the students by adopting health-promoting policies and practices.

Bearing this in mind, the USAID-funded Building Healthy Cities (BHC) project is providing technical assistance to Indore Smart City Development Limited (ISCDL) to implement a comprehensive HPS programme. This programme will promote the physical, social, emotional, mental and educational growth of the students by facilitating health-promoting policies and practices. The programme is being implemented by BHC with support from MP Voluntary Health Association (MPVHA), a leading NGO.

Implementation began in October 2019 in 148 government (aka public) middle, high and higher secondary schools of Indore selected by the Education Department. Out of 148 schools, 87 were co-education schools, 37 girls' schools and 24 boys' schools (Fig. 1). The total number of middle schools and high & higher secondary schools was 104 and 44, respectively.

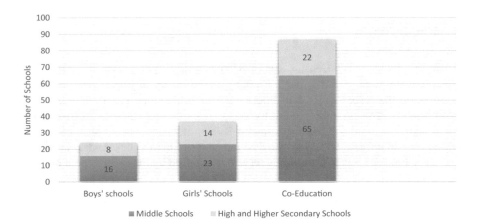

Fig. 1 Profile of selected schools

In developing the training resources for HPS, the objectives were:

(i)   to train teachers as health ambassadors with adequate knowledge and skills to promote healthy behaviour and lifestyle among students; and

(ii) to assess physical and social environment of schools and inform appropriate authorities for corrective action.

HPS trainings were conducted from November 2019 to January 2020 in 6 batches with 50 teachers in each batch. Trainings were conducted by subject experts using the Teachers Training Manual developed by BHC which focuses on 8 themes detailed in the following section.

Schools will be assessed pre- and post-training using WHO's HPS assessment tool. Results of the assessment will be reported once post-training results are finalized. The schools are assessed on their physical infrastructure, compliance to health policies, healthy social environment, engagement of parents and community, healthcare and promotion services. If schools qualify in the post-training assessment, they will be certified as Health-Promoting Schools.

3   Findings

The design and implementation of HPS in Indore relies on three key factors to create positive change. They are training of teachers as health ambassadors, promoting healthy school environment and multisectoral collaboration and convergence. While further exploration of their impact will occur during the assessment, below we define these factors, how they came to be and how they were implemented.

3.1   Training of Teachers as Health Ambassadors

Understanding the importance of school health education, two teachers each (preferably the principal and science teacher) from each school were trained by subject experts using the Teachers Training Manual prepared by the BHC team. This manual provides age-appropriate guidelines for all stakeholders including students, teachers and parents with the goal to involve the students in making positive and healthy lifestyle choices through systemic interventions.

The manual focuses on the following themes:

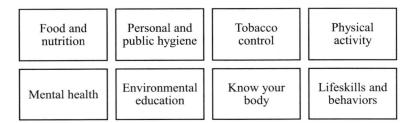

Food and nutrition	Personal and public hygiene	Tobacco control	Physical activity
Mental health	Environmental education	Know your body	Lifeskills and behaviors

The implementation of training relied on the buy-in and enthusiasm of the individual teachers. The two-day training was conducted in 6 batches; each batch comprised of 50 teachers. The subject experts used various training techniques like presentation-discussions, audio-visual aids, group activities, etc. The trainings capitalized on teachers' pedagogical knowledge, knowledge of student's social and health context, major issues that they face, their experiences and skills. These ideas were used to further enrich the training manual.

Teachers gave extremely positive feedback during the training sessions. They credited the acquisition of HPS in the curriculum and stated that it is essential to make the school environment better for the students. Teachers affirmed that these trainings will prove very helpful in educating the students and in implementing health policies in schools. Based on the response, BHC is planning additional outreach to keep these "Health Ambassadors" engaged with updated information.

3.2 Promoting Healthy School Environment

In addition to the HPS training, BHC assessed physical environments in the participating schools. No matter how well-trained teachers and students may be, the physical environment must also support healthy behaviours in order for this approach to succeed. The pre-assessment found that the majority of schools had sufficient ventilation, light and space as well as safe water and separate toilets for boys and girls (Fig. 2). However, many schools lacked physical and human resources relating to physical activity (playgrounds, sports materials and sports teachers) and for health support (sanitary napkins, deworming and health records).

In order to implement positive changes to the physical environment, BHC and its local partners are sharing these results with the relevant city authorities, as well as including these results in larger BHC efforts to inform municipal work-planning and funding priorities.

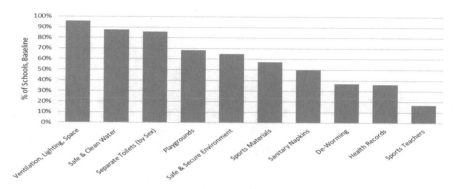

Fig. 2 Baseline performance on physical aspects of HPS

3.3 Multisector Collaboration and Convergence

Finally, HPS also focuses on bringing together various departments to develop healthy learning spaces for students. The objective is to help stakeholders address issues in a focused manner, pool resources and formulate common objectives. One of its major advantages of this multisector approach is the optimization of resources by avoiding duplication of inputs and activities. To make this easier, BHC supported the formation of a "multi-sector smart health working group" in Indore which was formally convened and is chaired by Chief Executive Officer, ISCDL and co-chaired by the Chief Medical and Health Officer. Figure 3 depicts the multiple stakeholders involved in implementing the activity in Indore. In practice, this working group covers all aspects of making Indore a healthy, livable Smart City for all. It will be important to ensure that any plans for school upgrades are reconciled with other infrastructure plans, and consider non-traditional funding options to ensure the sustainability of the HPS approach for many years to come.

4 Policy Recommendations

Indore is off to a strong start to creating a set of health-promoting schools at the middle and high school level. In order to build on this momentum, BHC suggests the following additional steps:

1. *Develop a supportive local government policy for HPS.* Though supportive national policies for HPS are in place (e.g. the School Health Programme), a local policy should be made and the HPS concept adopted by all schools including government schools which were not included in the pilot, and private ones.
2. *Gather support from the Department of Education and school administration to ensure that HPS is a whole of school approach.* Support and commitment is needed from school teachers, principal and administrators in order to succeed.

Fig. 3 Multiple stakeholders
involved in the
implementation

3. *Set up a school health advisory committee.* To coordinate and monitor all health promotional efforts in the school including development of health-promoting school policies, members of the advisory council can include the head of the school, representatives from parents, teachers and students.
4. *Develop a local HPS charter.* This agreement will symbolize the city's commitment to develop schools as healthy spaces for students.
5. *Scale up to regional level.* The HPS manual developed by BHC can be introduced in schools at the regional level.

References

Government of India (n.d.) Provisional population totals: urban agglomerations and cities. Government of India. Accessed 10 Mar 2020

Kaur J, Sushma KS, Bhavneet B, Surinder K (2015) Health promotion facilities in schools: who health promoting schools initiative. Nurs Midwifery Res J 11(3)

Sankhe S, Ireena V, Richard D, Ajit M et al (2010) India's urban awakening: building inclusive cities, sustaining economic growth. Mckinsey Global Institute

World Health Organization (n.d.) who I what is a health promoting school? Accessed 16 Mar 2020

Mobility and COVID-19: Time for a Mobility Paradigm Shift

Carolyn Daher, Sarah Koch, Manel Ferri, Guillem Vich, Maria Foraster, Glòria Carrasco, Sasha Khomenko, Sergio Baraibar, Laura Hidalgo, and Mark Nieuwenhuijsen

1 Key Messages

Urban mobility and the COVID-19 pandemic have had significant impacts on each other and on health. Urban areas are particularly hit by COVID-19, where moving around while social distancing is challenging because public space is often limited and mostly designated for motorized traffic. Urgent actions need to be taken by

C. Daher (✉) · S. Koch · M. Ferri · G. Vich · M. Foraster · G. Carrasco · S. Khomenko · S. Baraibar · L. Hidalgo · M. Nieuwenhuijsen
Barcelona Institute for Global Health (ISGlobal), Barcelona, Spain
e-mail: carolyn.daher@isglobal.org

Universitat Pompeu Fabra (UPF), Barcelona, Spain

CIBER Epidemiología y Salud Pública (CIBERESP), Barcelona, Spain

S. Koch
e-mail: sarah.koch@isglobal.org

M. Ferri
e-mail: mferri1960@gmail.com

G. Vich
e-mail: guillem.vich@isglobal.org

M. Foraster
e-mail: maria.foraster@isglobal.org

G. Carrasco
e-mail: gloria.carrasco@isglobal.org

S. Khomenko
e-mail: sasha.khomenko@isglobal.org

S. Baraibar
e-mail: sergio.baraibar@isglobal.org

© Zhejiang University Press 2021
F. W. Gatzweiler, *Urban Health and Wellbeing Programme*,
Urban Health and Wellbeing,
https://doi.org/10.1007/978-981-33-6036-5_6

cities and citizens that bring longer-term changes towards healthy, equitable and sustainable mobility.

1. The highest priority should be given to active transport and providing sufficient public space for movement, while maintaining physical distancing.
2. We need better use of technology to organize mobility and clear communication about the available mobility network options to alleviate fear and encourage rational transport choices.
3. Transport choices should be based on: (1) the risk of transmission, (2) health and environmental impacts and (3) access to and use of space.
4. We recommend:

   - Walking, cycling or Personal Mobility Vehicles for journeys up to 5 km.
   - Cycling for journeys up 10 km (and electric cycling for longer).
   - Low occupancy public transport for longer journeys.
   - Cars and motorcycles for vulnerable populations and those who cannot use the other transport modes.

2 Overview

Over the past decades, urbanization and transport infrastructure have drastically changed global landscape. Facilitating and managing mobility within cities and metropolitan areas already posed serious challenges for governments for decades, but has been substantially aggravated by the coronavirus (COVID-19) pandemic.

The relationship between health, mobility and climate has multiple dimensions (Nieuwenhuijsen 2020). Especially in urban areas, motorized traffic, particularly private vehicles, causes most air and noise pollution, the two main environmental health threats. Car-focused urban planning that gives the majority of public space to roads and parking contributes to the heat island effect, sedentary lifestyles and lack of green spaces. On the other hand, equitable access to safe mobility options is crucial for economic recovery, ensuring provision of and access to goods and services. Access to outdoor and natural spaces, where transmission risk is reduced, is critical to promoting population health.

In response to the pandemic, many countries have implemented drastic interventions to reduce COVID-19 transmissions, including varying forms of confinement, teleworking and travel restrictions. In the acute phase, single occupancy car and motorcycle use in cities fell, and thereby also air pollution and noise levels. Active transportation such as cycling and walking increased dramatically in some countries, while public transport collapsed in many cities amidst fear of greater transmission

L. Hidalgo
e-mail: laura.hidalgo@isglobal.org

M. Nieuwenhuijsen
e-mail: mark.nieuwenhuijsen@isglobal.org

risk, resulting in severe economic impacts on this sector. These acute changes have led governments and citizens to rethink their approach to urban mobility.

COVID-19 adaptations and distancing measures will stay in place for the foreseeable future. How we can keep private motorized vehicle use low, restore confidence in public transport use and promote active mobility for a sustainable, equitable, habitable and healthy society after the pandemic?

3 What Are the Main Mobility-Related Health Impacts Related to COVID-19?

3.1 Air Pollution

Motorized traffic in cities is the main contributor to air pollution, especially particulate matter ≤ 2.5 mm ($PM_{2.5}$) and nitrogen dioxide (NO_2). Worldwide, outdoor air pollution causes over four million deaths each year. Air pollution is also starting to be linked with the COVID-19 disease. Preliminary evidence suggests higher viral spread, development of more severe COVID-19 disease progressions and increased COVID-19-related mortality in more polluted areas (Ogen 2020; Setti et al. 2020; Wu et al. 2020).

Air pollution levels have dropped in many cities worldwide due to the confinement measures, lower economic activity and decrease in mobility. However, long-term reductions are needed to achieve significant health benefits (Barcelona Institute for Global Health 2020). The current crisis shows that major reductions in air pollution levels are possible and offers a critical opportunity to make lasting positive health impacts through more active and sustainable urban mobility solutions. Changes in infrastructure and policy to facilitate and incentivize active and sustainable transport are urgently needed to avoid a pollution rebound effect.

3.2 Noise

Regular exposure to environmental noise contributes to persistent stress and annoyance, sleep disturbance, and in the long term leads to chronic conditions such as cardiovascular disease and diabetes (Van Kempen et al. 2018). Road traffic is usually the main source of noise in many cities. COVID-19 confinement has led to an enormous decrease in noise in cities worldwide: Paris has recorded a 90% reduction in some streets; in Madrid and Barcelona, levels dropped massively by 13 and 11 decibels, respectively, compared with the 2019 average.

Quieter cities are possible; however, short-term reductions are not sufficient to improve health. Cities need to reinforce plans for changes in mobility patterns to maintain longer-term lower noise levels.

3.3 Physical Activity

Urban transport planning influences citizens' physical activity. A sedentary lifestyle is the fourth risk factor for global mortality and is associated with 6% of deaths worldwide (World Health Organization 2018). Enabling physical activity is even more urgently needed during the pandemic to minimize negative health impacts. Among many health benefits from physical activity are an enhanced immune system and a stabilization effect on mental health, two aspects that have been shown to be particularly affected by COVID-19 and the associated confinement restrictions (ISGlobal[1] 2020; ISGlobal[2] 2020). It is crucial that urban residents have options for physical activity through active mobility and access to outdoor spaces and natural areas (Fig. 1).

4 How Can Mobility Contribute to COVID-19 Management and Beyond?

Urban and mobility planners need to protect public health by inverting the current transport pyramid and facilitating mobility that allows people to meet their basic needs in the safest possible way. Cities can lead the way in developing strategies through social, tactical and technological policies and interventions. This requires efficient and effective collaboration across sectors and in conjunction with society. Measures should be implemented rapidly, but with the intent to create long-term positive changes.

Fig. 1 Cycling as a form of active transportation resulting in increased physical activity

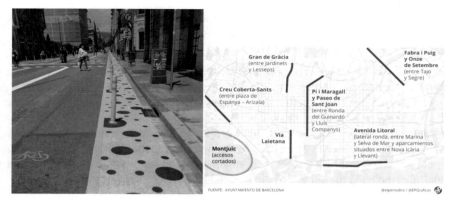

Fig. 2 Examples of street interventions to facilitate active mobility in Barcelona, Spain

4.1 Reallocate Public Space to Prioritize Active and Local Mobility Whenever Possible

We should take advantage of the 60–90% fewer motorized trips that have freed up considerable public space in the acute COVID-19 pandemic phase to prioritize active mobility and devote large areas for bicycles and pedestrians that would allow new users to be safely added to the group of active commuters, and a larger number of trips among current users (Fig. 2).

Walking or cycling are the two healthiest, most sustainable and equitable options which guarantee social distancing. Scooters and other personal mobility vehicles (PMVs) will also become more important. These offer opportunities to decongest transport, thus facilitating compliance with distancing. Many cities have already begun to enable road infrastructure to promote active transport (Polis Network 2020). Local trips should be prioritized whenever possible for daily activities and services, thus incentivizing active mobility as main transport modes.

4.2 Make Public Transport as Safe as Possible

For many, including essential workers, public transport is the only viable option for daily mobility. Vulnerable populations that rely on public transport will also be disproportionately affected by disruptions. Local governments and traffic authorities must work together to provide adequate service while maintaining safe conditions. Reducing overcrowding by increasing public transport and/or controlling the number of passengers, improving the ventilation, frequent disinfecting and requiring the use of masks are important measures to reduce the risk of transmission. Encouraging mixed-mode transport, such as combining bicycles or PMVs with public transport, can help decrease crowding at the beginning and end of longer trips.

4.3 Promotion of the Rational Use of Private Transport, Taxis and Shared Vehicle Services

Use of private vehicles may help keep distance, reducing the risk of transmission, but comes at the cost of high emissions, noise and urban space occupancy. Further, cities with high traffic volumes before the pandemic will struggle to manage an increase in circulating vehicles, and this will limit the space for active transport. Taxis and shared vehicle services offer more flexible options for people who may need to use cars, especially for more vulnerable populations, such as the elderly. If implemented in conjunction with city transport management, they are a viable option that can minimize the need for individual car ownership.

4.4 Use Technology to Manage and Programme Mobility

Technology is a key asset for mobility management that is underused. Smartphone applications can help people find optimal routes and suggest alternatives to avoid crowding. For instance, apps can alert people about congested streets in real time and can be used to plan trips on public transport in advance, to limit occupancy. They can also be used for payments to avoid having to touch ticket machines in stations and on buses. Technology tools can also be used to create passenger flow and alert systems for transport operators.

4.5 Change Working and Shopping Habits

Providing more flexible options for teleworking and managing work-time hours will help decongest routes. Planning strategies, such as the 15-minute city, encourage local shopping to reduce trips and encourage those accessible by active. Online shopping can reduce individual trips; however, delivery of goods should be coordinated and preferably done through active and sustainable transport during the "last mile" (Fig. 3).

4.6 Immediate Actions Recommended

Linking health and mobility at this time of transformation can bring large gains and move towards achieving the Sustainable Development Goals. Applying measures to increase active transport, in combination with public space management to guarantee distancing, will assist in combating the spread of COVID-19. These strategies also promote a healthier lifestyle, while reducing the impacts on the environment. However, changes need to be consolidated in the longer term and carefully monitored.

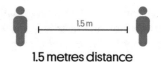

Mobility options

1.5 metres distance

Impact on use of public space

	CR	Space	Health benefits	Environ. impacts
Cars	L	H	L	H
Motorbikes	L	M	L	H
Public transport	?	M	M	M
Walking	L	L	H	L
Cycling	L	L	H	L
Other, incl. PMV	?	?	?	?

Impact:

⬜ positive   ⬛ negative   ▨ intermediate   ▨ unknown

CR = contagion risk
PMV = Personal Mobility Vehicle
L = low; M = medium; H = high

ISGlobal Barcelona Institute for Global Health

Fig. 3 Mobility options including their contagion risk, space, health benefits and environmental impact

Below is a selection of measures being implemented in different cities worldwide. Organizations such as the WHO, national and local transport agencies, have made available more extensive guidelines for safe mobility during the pandemic.

4.7 Public Space and Active Transport

- Promote active transport for short trips
- Close or reduce car lanes on streets with two lanes (or more) each way to accommodate cycle paths and wider pavements for pedestrians
- Close streets to motor vehicles in more residential areas to open them for pedestrians and cyclists.

4.8 Public Transport: Bus, Underground and Railway

- Expand the frequencies and limit occupancy, especially during the rush hours
- Increase cleaning and disinfection and distribute masks
- Implement traffic light priority for buses and segregated lanes to increase speed and passing times.

4.9 Private Transport

- Reduce parking space for private vehicle
- Implement low emission and 30 km/hour zones.

Link taxi and shared vehicle transportation with the organization of public transport.

References

Barcelona Institute for Global Health (2020) Are the reductions in air pollution triggered by the COVID-19 epidemic having health effects? [2020-08-12]. https://www.isglobal.org/en/healthisg lobal/-/custom-blog-portlet/-sera-relevante-para-nuestra-salud-la-disminucion-de-la-contamina cion-atmosferica-durante-la-epidemia-de-la-covid-19-/5083982/11101

Isglobal[1] (2020) ¿Debería permitirse la actividad física durante la pandemia causada por el coronavirus? [2020-08-12]. https://www.isglobal.org/es/-/-deberia-permitirse-la-actividad-fisica-dur ante-la-pandemia-causada-por-el-coronavirus

Isglobal[2] (2020) ¿Deberíamos salir durante y después del confinamiento por la COVID-19? [2020-08-12]. https://www.isglobal.org/es/-/-deberiamos-salir-durante-y-despues-del-confinami ento-por-la-covid-19-

Nieuwenhuijsen MJ (2020) Urban and transport planning pathways to carbon neutral, liveable and healthy cities; A review of the current evidence. Environ Int 140:105661. [2020-08-12]. https://doi.org/10.1016/j.envint.2020.105661

Ogen Y (2020) Assessing nitrogen dioxide (NO_2) levels as a contributing factor to coronavirus (COVID-19) fatality. Sci Total Environ 726:138605. [2020-08-12]. https://doi.org/10.1016/j.sci totenv.2020.138605

Polis Network (2020) COVID-19: Keeping things moving—polis network. [2020-08-12]. https://www.polisnetwork.eu/document/covid-19-keeping-things-moving/

Setti L, Passarini F, de Gennaro G et al (2020) Evaluation of the potential relationship between Particulate Matter (PM) pollution and COVID-19 infection spread in Italy. SIMA position paper [2020-08-12]. http://www.simaonlus.it/wpsima/wp-content/uploads/2020/03/COVID_19_ position-paper_ENG.pdf

Van Kempen E, Casas M, Pershagen G et al (2018) Who environmental noise guidelines for the european region: a systematic review on environmental noise and cardiovascular and metabolic effects: a summary. Int J Environ Res Public Health 15(2):1–59. [2020-08-12]. https://doi.org/10.3390/ijerph15020379

World Health Organization (2018) Physical activity. Fact sheet. [2020-08-12]. https://www.who.int/news-room/fact-sheets/detail/physical-activity

Wu X, Nethery RC, Sabath BM et al. (2020) Exposure to air pollution and COVID-19 mortality in the United States. MedRxiv [2020-08-12]. https://doi.org/10.1101/2020.04.05.20054502

Find Out More

- Catalogue of the European Cyclists' Federation: https://ecf.com/cycling-beyond-crisis
- City of Bogota: https://bogota.gov.co/mi-ciudad/movilidad/distrito-estudia-hacer-permanentes-ciclovias-de-cuarentena-en-bogota
- CityLab Transportation: https://www.citylab.com/transportation/
- Emergency Mobility Network: Action Plan for Post COVID-19 Mobility: https://www.bikeitalia.it/wp-content/uploads/2020/04/RME-Piano-di-azione-mobilit%C3%A0-urbana-post-covid.pdf
- POLIS-Cities and Regions for Transport Innovation: https://www.polisnetwork.eu/document/covid-19-keeping-things-moving/, https://www.polisnetwork.eu/document/resources-covid-19-mobility/
- WHO Moving around during the COVID-19 outbreak: http://www.euro.who.int/en/health-topics/health-emergencies/coronavirus-covid-19/novel-coronavirus-2019-ncov-technical-guidance/coronavirus-disease-covid-19-outbreak-technical-guidance-europe/moving-around-during-the-covid-19-outbreak

COVID-19 Shows Us the Need to Plan Urban Green Spaces More Systemically for Urban Health and Wellbeing

Jieling Liu

1 Key Messages

In addition to the necessity and urgency for bottom-up climate actions, the significance of urban green spaces has been highlighted by the complex socio-economic impacts of COVID-19 on a global scale.

1. Sustainable urban development needs to go beyond segmented actions and to take a more systemic approach to urban planning, particularly that of green spaces.
2. Urban green spaces are particularly invaluable for those who cannot afford a spacious living environment, such as migrant workers and other vulnerable groups.
3. Urban green spaces should be treated as an invaluable common-pool resource for common health and wellbeing, i.e. one's use of green spaces should not reduce the availability of them for others, yet it is impossible and inadequate to exclude any residents from using green spaces regardless of their socio-economic status.
4. Policymakers should establish universal access to urban green space as the basic principle of urban planning, integrate climate modelling and public health monitoring into green space planning, and incorporate all societal actors in urban green space governance from planning to building to use and further to maintenance, for enhancing both social and environmental sustainability.

J. Liu (✉)
Climate Change and Sustainable Development Policies at Institute of
Social Sciences, University of Lisbon, Lisbon, Portugal
e-mail: jielingliu@campus.ul.pt

© Zhejiang University Press 2021
F. W. Gatzweiler, *Urban Health and Wellbeing Programme*,
Urban Health and Wellbeing,
https://doi.org/10.1007/978-981-33-6036-5_7

2 Background

Many countries and societies have approximated sustainable development through reducing carbon emissions, e.g. by applying green technology in transportation and in industry, reduction in deforestation and in plastic consumption. Yet, the current COVID-19 pandemic has (repeatedly, if we think of historical pandemics) made it clear to all of us that health is an important common denominator in any issues or risks urgently needed to be addressed today, from air pollution to plastic abuse, to income inequality, to the lack of essential live supporting infrastructure. An overview of human history tells us that health has always been the ultimate goal of all human beings and literally speaking all kinds of human collaborations we have come to devote ourselves to, including the current paradigm to pursue sustainable urban development, serve to achieve this goal. The global implications of COVID-19 on all sectors of socio-economic activities have demonstrated clearly that, much of the sustainable urban development practices are still either siloed, compartmentalized or too short-term, lacking synergy between one activity to another. Sustainable urban development needs to go way beyond these segmented actions and to take a more systemic approach to urban planning, particularly that of public and/or green spaces. The reasons are the following:

First, the irreversible urbanization trend and increasing urban density. With more than 50% of the world population living in cities, we are already living on an urban planet (Bai et al. 2018, p. 462). Yet, urban areas account roughly for only 3% of the earth's surface (Ryan 2014; Liu et al. 2014). With increasing urbanization come several consequences. Urban areas where the majority of the population inhabit today are increasingly dense. The dense population implies increasing condensation of artificial infrastructure including transportation, housing, commercial, sanitation and other services; it also means a drastic reduction in ecological or green infrastructure such as parks, ecological corridors, wetlands, lawns, and other natural landscapes. For most of human history, populations had lived within small-scale and low-density social groups; urbanization is a trend unique to the past few centuries (Ryan 2019; Ritchie and Roser 2020).

The particular high mortality rate among the African American communities across US cities during COVID-19 has shown how dense living environments and the implied underprivileged socio-economic conditions are jeopardizing human health. African Americans have denser and older housing in many cities in the United States owing to the historical legacy of "redlining", a zoning policy made by the Federal Housing Administration to demarcate African American neighbourhoods initially to support affordable housing but ended up as a barrier to housing mortgages (Gross 2017; Vock et al. 2019). An example of the COVID-19 impact hitting these communities particularly hard: African Americans in Louisiana make up 70% of the COVID-19 deaths in all known coronavirus patients in the state, while only representing about 32% of the population (Turk 2020).

The second reason is the global interconnectivity. Ever since the old Silk Roads emerged beginning in the first century BC when Chinese luxury products started to appear in Rome, economic activities have begun to become global. Past industrial revolutions continued to interconnect raw materials, talents, services from all around the world. With the complex, hyperconnected globalized economic system today, one can hardly survive on local produce. The free-market economic system, based on which much of the urban areas have thrived, have used the convenience of globalized technology, international financial institutions and regional trade agreements to take advantage of abundant, cheap labour in the developing countries. The results have degraded the environmental quality in many emerging urban areas and offered pittances of economic benefits to labourers who are often referred to as urban migrants.

On the one hand, the high dependency of global interconnectivity for daily life in the urban era has drastically accelerated the spread of diseases such as COVID-19, compared to the pandemic of flu a century ago. On the other hand, increased air pollution, global climate change and the accompanying extreme weather events have increased the challenge of maintaining healthy lifestyles in dense and highly interconnected urban communities worldwide, weakening their ability to cope with epidemic diseases.

3 Insights and Findings

In addition to the necessity and urgency for bottom-up climate actions at the city level, the significance of urban green space resources has been further highlighted by the complex socio-economic impacts of COVID-19 on a global scale. As space in cities is increasingly claimed, marked, planned, and priced meticulously, public open spaces, such as parks, green corridors, lawns, are particularly invaluable spaces for the urban population, particularly for those who cannot afford a spacious living environment, such as migrant workers and other vulnerable groups. Open, walking spaces can contribute to enlarge distances between pedestrians, reducing the risk of transmission due to the dependency on public transportation, and furthermore, encourages walking and biking, which helps to maintain health. Google's global COVID-19 community mobility data shows that global mobility trends for park spaces during the quarantine period (mid-February–mid-May) decreased about 4.5% compared to pre-corona baseline; while mobility trends for retail and recreation spaces decreased around 22% and for workplaces decreased around 24% compared to baseline. These numbers highlighted the importance of urban green spaces. Sightings of nature during COVID-19 have seemed to help many people gain an alternative perspective to the one that had long been human-centric.

Urban green spaces in cities should be treated as an invaluable common-pool resource for our common health and wellbeing for two main reasons (Ostrom 2009). First, no one should be excluded from using urban green spaces for health and wellbeing purposes; technically speaking, it is also impossible to exclude any urban residents from benefiting from the multiple ecosystem services provided by urban green spaces, e.g. purified air and groundwater, and greater biodiversity. Second, urban green spaces in cities are limited due to the scarcity nature of urban land. That means one resident's use in existing green space will reduce the availability of that space for others; it also means one decision of land planning and land use for commercial or housing purposes will reduce the opportunities for additional green spaces. These two attributes determine that the current approaches of urban green space planning which rely on narrow indicators of success—diversification of tree species, coverage (certain size of green space per capita), aesthetics, etc., are insufficient. If a city wants to retain considerable urban green spaces for climate adaptation and make it accessible to everyone for wellbeing benefits, some form of integrative institutional arrangements for planning and governing urban green spaces elaborating all urban functions as a complex adaptive system must be introduced.

4 Policy Recommendations

Urban spatial planning and social policymaking share the role of distributing spaces more justly, taking into account that public spaces should first and foremost be able to serve all equally regardless of socio-economic status. Applying a systemic approach to plan and govern urban green space can help improve urban health and wellbeing. Policy recommendations for decision-makers at the city level include:

1. Establish universal access to urban green space as the basic principle of urban planning.
2. Integrate the modelling work of climate change impact and vulnerability and the monitoring system of public health into the planning for urban green spaces.
3. Incorporate all societal actors in the governance of urban green spaces from planning to building to use and to maintain, taking into account the ecology of plant species.

References

Bai X, Elmqvist T, Frantzeskaki N et al (2018) New integrated urban knowledge for the cities we want. In: Elmqvist T, Bai X, Frantzeskaki N et al (eds) Urban planet: knowledge towards sustainable cities. Cambridge University Press, Cambridge.

Gross T (2017) A 'forgotten history' of how the U.S. government segregated America. [2020-08-12]. https://www.npr.org/2017/05/03/526655831/a-forgotten-history-of-how-the-u-s-government-segregated-america.

Liu Z, He C, Zhou Y et al (2014) How much of the world's land has been urbanized, really? A hierarchical framework for avoiding confusion. Landscape Ecol 29:5

Ostrom E (2009) Understanding institutional diversity. Princeton University Press, Princeton

Ritchie H, Roser M (2020) Urbanization. [2020-08-12]. https://ourworldindata.org/urbanization

Ryan BJ (2014) Foreword: GEO—a globally integrated approach to urban monitoring. In: Weng Q (ed) Global urban monitoring and assessment through earth observation, remote sensing applications series. CRC Press.

Ryan C (2019) Civilized to death. Avid Reader Press/Simon & Schuster, New York

Turk S (2020) Racial disparities in Louisiana's COVID-19 death rate reflect systemic problems. [2020-08-12]. https://www.wwltv.com/article/news/health/coronavirus/racial-disparities-in-lou isianas-covid-19-death-rate-reflect-systemic-problems/289-bd36c4b1-1bdf-4d07-baad-6c3d20 7172f2

Vock D, Brian C, Mike M (2019) How housing policies keep white neighborhoods so white (and black neighborhoods so black)—decades of local zoning regulations and land-use policies have kept racial segregation firmly rooted in place. [2020-08-12]. https://www.governing.com/topics/ health-human-services/gov-segregation-housing.html

How Lack or Insufficient Provision of Water and Sanitation Impacts Women's Health Working in the Informal Sector: Experiences from West and Central Africa

H. Blaise Nguendo Yongsi

1 Key Messages

1. Like other African countries, population of Sub-Saharan Africa is characterized by a relatively high proportion of women (more than 50%). This female population constitutes an undeniable workforce in the economic development process of those countries.
2. Across Sub-Saharan Africa, women are, to varying degrees, at the forefront of the economic life. Due to the low economic development level of those countries and mostly of sociocultural constraints, they mainly operate in the informal service sector such as street catering, retail sale, cleaning and maid service.
3. However, their working environment is less or poorly provided with water and sanitation infrastructure. This poses a serious problem of sexual/reproductive health. In fact, management of their menstruation becomes a challenging situation as they don't have access to water and sanitation.
4. As a matter of consequence, they can no longer carry out their activities during a time period, which represents a significant socio-economic loss. It is therefore crucial to involve women in the design of improved sanitation and waste management structures.

2 Water and Sanitation Impacts on Women Health

At least 500 million women and girls globally lack adequate facilities for menstrual hygiene management. Inadequate WASH (water, sanitation and hygiene) facilities,

H. B. N. Yongsi (✉)
Urban Health at IFORD-University of Yaoundé II, Yaoundé, Cameroon

Catholic University of Central Africa, The School of Health Sciences, Yaoundé, Cameroon

© Zhejiang University Press 2021
F. W. Gatzweiler, *Urban Health and Wellbeing Programme*,
Urban Health and Wellbeing,
https://doi.org/10.1007/978-981-33-6036-5_8

particularly in public places, such as in schools, workplaces or health centres, pose a major obstacle to women and girls. The lack of safety toilets, and the unavailability of means to dispose of used sanitary pads and water to wash hands, expose women and girls to face challenges in maintaining their menstrual hygiene (i.e. their sexual health) in a private, safe and dignified manner. A growing body of evidence shows that women and girls' inability to manage their menstrual hygiene in schools and in working environment, results in school and work absenteeism, which in turn, has severe economic costs on their lives and on the country (ANSD 2016; INS/ICF International 2012).

Results presented here derive from a joint WSSCC/UN women research under the framework of the Joint Programme on Gender, Hygiene and Sanitation implemented in 3 countries of West and Central Africa (Cameroon, Niger and Senegal). Data were collected using a cross-sectional epidemiological design and a mixed quantitative and qualitative analysis. Our work has led to two main results:

2.1 Less Provision, Unsatisfactory and Unsafe Wash Infrastructure in Women Public Working Places

In Sub-Saharan African countries, access to WASH facilities is a challenge for women working in public places. Findings show that regardless of the country, toilets used by women are dominated by the latrines, moreover insufficient because the WHO standard of 20 individuals per latrine is rarely applied (Fig. 1).

Visited facilities are poorly managed and irregularly cleaned (twice per week). Maintenance here relies only on little repairs (Fig. 2).

Not only those toilets are in poor condition, but they are also unsafe. All countries taken together, the toilets display the following characteristics: limited space, no locks on doors (when/where it exists), no lighting (Fig. 3). In short, more toilets need as well as better maintenance. And with regard to menstrual waste management, it appears to be an overlooked issue.

2.2 Impact on Women's Health and Economic Activities

A high proportion of women claim that the onset of menstruation is a time of pyscho-logical and physical ill-being, because of menstrual-related health problems they regularly face (Table 1).

Always in relation to their daily life, women assert to be excluded and stigmatized within their family and community during menstrual periods. In Cameroon, 44 and 56% of women have reported having suffered, respectively, from isolation and stress along with stigmatization. In Niger, they represented 39 and 61%, whereas in Senegal, they were 47 and 53% (Mitullah et al. 2016; UNICEF 2015).

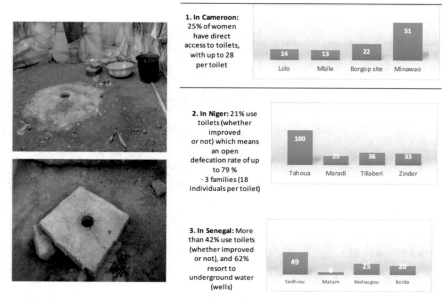

Fig. 1 Level of use of Latrines

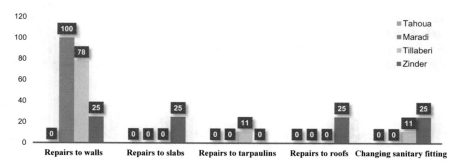

Fig. 2 Management on visited facilities

With regard to the impact of poor access to WASH facilities on women's socio-economic activities, our findings show that during their menstruations and due to lack of/unsafe toilets:

Though most domains (domestic, administrative, economic, school activities) are impacted by women inactivity during their menstruation, slowdown is more much felt in the economic sphere, all the more so because they are involved in the economic development of their households and community.

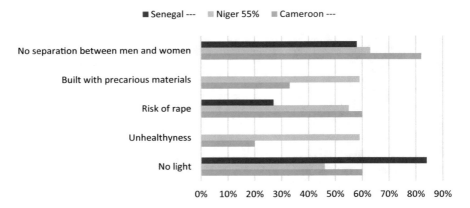

Fig. 3 Safety concerns

3 Conclusion

The challenge menstruating girls and women face is often less tangible than simply the availability of infrastructure, and is rooted in social norms and beliefs (United Nations 2015). Given the multiple challenges women and adolescent girls face, it is evident that promoting menstrual hygiene management is not only a sanitation matter; it is also an important step towards safeguarding the dignity, bodily integrity and overall life opportunities of women. Through a range of technical and analytical initiatives, our Research Laboratory aims to better understand and address issues of management of hygiene menstruation and to thereby elevate the dialogue on its importance.

4 Recommendations

1. Increase the amount of WASH infrastructure in communities and households, and strengthen measures for the cleaning and maintenance of existing infrastructure.
2. Develop the construction of gender-separated latrines in educational establishments and public places, especially those frequented by economically active women, most of whom work in the informal sector.
3. Raise awareness of the negative consequences of poor waste management on health and the environment.
4. Develop waste management systems at the local level and facilitate access to water, sanitation and waste disposal for women and girls, especially in rural areas.

Table 1 Percentage of women with menstrual-related health problems

	Menstrual-related health problems									
	General fatigue	Vertigo and nausea	Headache	Itching and pimples around the vaginal area	Fever	Cough and cold	Backpain	Painfulbreasts	Stress	Malaria
Cameroon	43	26	25	13	17	03	42	24	13	7
Niger	72	11	24	17	56	38	9	5	52	43
Senegal	—	38	49	8	30	4	56	43	—	6

5. Involve women in the design of sanitation and waste management structures to
    ensure menstrual waste-related needs for disposal, collection and treatment are
    taken into account.

References

ANSD (2016) L'Enquête Nationale sur l'Emploi au Sénégal. Dakar, p 114

INS/ICF International (2012) Enquête démographique de santé Cameroun 2011 [en ligne]. Yaoundé:
    Institut national de la statistique / ICF International. [2020–08–12]. https://dhsprogram.com/pubs/
    pdf/FR260/FR260.pdf

Mitullah W, Samson R, Wambua P et al (2016) Malgré un certain progrès, les infras-
    tructures de base demeurent un défi en Afrique. Dépêche N° 69 d'Afrobaromètre.
    [2020–08–12]. https://afrobarometer.org/fr/publications/ad69-malgre-un-certain-progres-lesinf
    rastructures-de-base-demeurent-un-defi-en-afrique

UNICEF (2015) Quelques faits et chiffres sur la situation des femmes au Niger. WCARO-Niamey,
    p 2

United Nations (2015) Goal 6: ensure access to water and sanitation for all. Sustainable development
    goals [2020–08–12]. https://www.un.org/sustainabledevelopment/water-andsanitation/

Planning Models for Small Towns in Tanzania

Dawah Lulu Magembe-Mushi and Ally Namangaya

1 Key Messages

1. In Tanzania, a large population with vibrant economic activities mostly informal, is found within cities, regions and small towns. These small towns are growing rapidly and strongly associated with mismanagement especially in their infancy stages.
2. The impact of mismanagement is manifested in expanding informality, both in economic activities as well as housing and social service provision.
3. Spatial informality implies inadequate provision of basic utilities and social services, horizontal congestion of people and activities and inability to geocode residences and functions.
4. These characteristics fasten the spread of communicable diseases like COVID-19 and the approaches of contact tracing which towns are relying upon to curb the spread of the pandemics become almost inapplicable.
5. Small towns require special attention in making sure that, the spread of such diseases is controlled through better planning, improved servicing and better development control and urban management.

D. L. Magembe-Mushi (✉) · A. Namangaya
School of Spatial Planning and Social
Sciences, Ardhi University, Dar es Salaam, Tanzania
e-mail: dawah.mushi@aru.ac.tz

A. Namangaya
e-mail: ally.namangaya@aru.ac.tz

© Zhejiang University Press 2021
F. W. Gatzweiler, *Urban Health and Wellbeing Programme*,
Urban Health and Wellbeing,
https://doi.org/10.1007/978-981-33-6036-5_9

2 Background

Capacity gaps and institutional conflicts in small towns are increasingly contributing
to unsustainable utilization of resources as well as the control of growth and management of small towns. This is manifested by expanding informality, land use conflicts,
degradation of resources and increased spatial and population growth. Following the
COVID-19 pandemic, these small towns have not slowed down their activities which
involve movement and interaction of people from different areas within and outside
the region and even the country. Considering high informality and scarcity of basic
facilities such as health centres, chances are these small towns can fuel the spread of
the virus. This is because it is very difficult to practise the recommended measures
for slowing down cases as well as containing the affected population. Due to lack
of planning, it is difficult to trace, isolate and contain the sick or practising social
distancing.

Our research applied case study methods within 25 non-regional urban centres
(see Fig. 1), which were studied in a historical perspective. Five of them (Geita,
Gairo, Handeni, Kabuku and Kibiti) were analysed in detail. The research explored
spatial transformation processes and drivers for changes of some village settlements
into urban centres. Physically the research explored land use organization of the

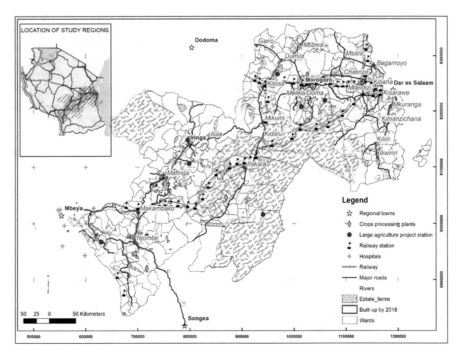

Fig. 1 Location of some of the case study small towns in their respective regions in Tanzania

settlements in different growth periods focusing on density changes, availability of common spaces and infrastructure way-leaves and influence of trunk infrastructure services on land use organization. On the governance side, the research explored interests, capacities and influences of different actors and stakeholders in planning process and management of land and activities in these emerging and growing small urban centres. These actors included landowners, local governments, local politicians and utilities service providers, business communities and civil society organizations. The research which employed action research not only focused on exploring maladies in the growth of the small towns and understanding prerequisites for integrating informal powers in land governance and improving services, but also piloted on ideas of re-configuration of the idea of proactive interventions in order to cope with informality, which is beneficial in emergency situations such as COVID-19. From this four-year engagement with communities in small towns, lessons were drawn that could improve the health and wellbeing of small towns and communities in the face of this and similar pandemics in the future.

3 Insights and Findings

It was observed that there is ambiguity and institutional discourse on what constitutes small towns, particularly between the concerned Ministries such as that dealing with Lands, Human Settlements Development (MLHHSD), and that of Regional Administration and Local Governments (PO-RALG) and National Bureau of Statistics (NBS) dealing with population and hierarchy of settlements within Tanzania. To exemplify, all three institutions use only population criteria in defining small towns but they differ in the applicable thresholds. Reference was made to the National Human Settlement Policy, statistical standards and World Bank definition of small towns (Muzzini and Lindeboom 2008). This affects their declaration, institutionalization (allocating responsible offices) planning and resourcing for development. Therefore, these observed ambiguities are also felt in dealing with the COVID-19 pandemic, where centralized decisions fail or are delayed in filtering into communities and data on the diseases spread and containment become unreliable.

From over 100 urban centres existing in Tanzania, less than 20% have General Planning Schemes/Master Plans and most of those which have them are urban centres which are Municipalities or cities. Among over 100 urban centres with a population between 50,000 and 150,000, less than 10 have Master Plans. Studies have further established that even for those with valid Master Plans, the implementation level in terms of spatial conformity between the plan and actual use is less than 30%. However, the implementation of the approved and surveyed detailed planning schemes seems to be higher (over 50%). General Planning Schemes seems to be affected by a lack of realistic consideration of land ownership and use claims, disconformity with interest of utility agencies and developers as well as challenges in legitimacy in the planning process; particularly the fact that the institutional structure doesn't make it mandatory for landowners, utility agencies and economic actors, to engage directly in decisions on the use of land rather the process relies on council experts and political representatives.

Most of the small towns have 90% of informal development. On the other hand, there is potential that is not taped in small towns due to the integrative nature of actors that are living together in small towns. They can make face-to-face contact and achieve social cohesion, which is more difficult in larger towns. For example, utility agencies are operating independently while utility committees like water and environmental are operating under the same governance of the District Executive Director (DED) or Town Executive Officer (TEO). Not realizing the importance of this integration has led most small towns to not plan interventions. Another impact is that sometimes less vibrant centres are selected as district administrative centres which then is a dis-service to the growing centres. Lack of planning in these areas makes it impossible for the provision of social services and to have location addresses for traceability of the affected individuals due to the absence of an address system and property location data in informal space and economy. The informality of economic activities cannot facilitate lockdown or less interactions between centres, centre-to-region and centres to other countries.

4 Key Policy Recommendations

As there is a preference to finance and implement the detailed plans rather than general planning schemes, it is recommended that institutional structure of planning should focus on instruments that directly impact on property boundaries, ownership and uses that will facilitate better housing condition in communities, i.e. detailed planning schemes that will take into consideration the provision of services, planning standards and regulations to facilitate healthy living in urban areas rather than emphasizing on the development of general planning schemes.

Communicative planning platforms are needed to positively influence planning process. Stakeholders with tangible and critical interests on land, i.e. landowners, property developers, utility providers and land-related investors in health centres, schools, markets should be engaged directly in planning process rather than being politically represented through councillors, street chairpersons or local government experts. These stakeholders will facilitate and speed up the implementation of detailed plans.

Involvement of private firms along with grassroot institutions influenced urban planning and resource mobilization for plan-making and implementation. When there are mutual beneficial interactions among private sectors, politicians and grassroot institutions, the community is willing to contribute financial resources to prepare planning schemes and even make sure its adherence during the actual implementation. In that case, there is a potential of private firms to mobilize resources for planning process in small towns with rapid growth, at early stages.

The institutional structure that allows overlap of mandates and subordination under DED or TEO, utility providers and planning actors in small towns should be used to facilitate resource mobilization for planning projects and their implementation.

Due to the unaffordability of large project implementations and sustainable service provisions in terms of costs, quality of services, infrastructures and expertise, clusterization of operations among proximate small towns to benefit from economies of scale and promote alternative technologies that are affordable by the small-town communities, is recommended.

For coordination and conformity among mandated urban governance institutions, the definition of small towns/urban settlements should consider three variables concurrently, i.e. population size, building/population density and percentage of labour force in off-farm employment. In Tanzania, using populations thresholds, small towns could be considered as areas with population between 12,000 and 50,000 and with building densities of between 10 and 30 buildings per hectare.

The most important areas for early intervention for urban growth in Tanzania should be in areas with vibrant commercialized agricultural activities that result in value additions and economic linkages; tarmacked road or any multimodal transport node including road junctions situated about 2-hours' drive from the next similar or bigger urban centre. The government should do pre-emptive planning intervention in such areas to guide their growth.

There should be special consideration of urban health-related aspects in small towns since they are the focal points for rural populations, which are the largest group in Tanzania. Having well-planned small towns will lead into growth of well-planned regional and cities of the future since these small towns grow and develop into cities.

Reference

Muzzini E, Lindeboom W (2008) The urban transition in Tanzania. Building the Empirical Base for Policy Dialogue, pp 1–166

Coping with Extreme Circumstances Through Community-Led Local Nature Interventions: A Science-Based Policy Analysis

Diana Benjumea and Agnieszka Olszewska-Guizzo

1 Key Messages

1. Community-led local nature interventions are coping strategies to extreme events created by low-income communities to help them sustain their health and wellbeing.
2. Nature interaction and community activism enact feelings of stewardship and leadership in the communities that lead to nature preservation.
3. Local nature intervention projects led by low-income communities are part of a secondary green network that could be acknowledged as part of the main city's urban green infrastructure.
4. More quality research is needed to understand the social dynamics that led to the creation of local nature intervention projects and their impact on the health and wellbeing of the communities.

2 Community-Led Local Nature Interventions

Over the last decade, the city of Medellin, Colombia has implemented diverse programmes to promote urban resilience and eco-urbanism within its most vulnerable communities. From 2007, the local government launched the mega-plan—known locally as the "Plan Maestro Area Centroriental"—to recover public spaces at the city's peripheral border and control the proliferation of illegal settlements found

D. Benjumea (✉)
Thammasat University, Khlong Nueng, Thailand
e-mail: d.benjumea@neurolandscape.org

A. Olszewska-Guizzo
National University of Singapore, Singapore, Singapore
e-mail: a.o.guizzo@neurolandscape.org

© Zhejiang University Press 2021
F. W. Gatzweiler, *Urban Health and Wellbeing Programme*,
Urban Health and Wellbeing,
https://doi.org/10.1007/978-981-33-6036-5_10

57

across mountainous landscapes. Included in the Plan Maestro was the green belt (GB), an initiative that attempted to connect the most relevant green nodes in Medellin through walking and cycling paths, leisure centres, and other green infrastructure (e.g. urban farming, community gardens, parks). These projects have integrated participatory design strategies (PDS), which are methods that allow community members to take part in the design and construction of the places, while also creating a sense of place attachment and responsibility towards the new-built spaces. Although these projects rendered Medellin to be recognized as one of the "greenest" and most resilient cities in the world, it has been suggested that the GB projects are part of a governmental agenda that beautify low-income neighbourhoods through mega green-projects to help attract middle- and upper-class tourists. In the process, low-income residents become dispossessed of their local environment directly affecting their social capital. In other words, the community organizations are disempowered, and the communities become dependent on the local government interventions.

An alternative to government PDS interventions is a process that has evolved over time inside the communities, known as community-led participation. The participative process of this approach occurs spontaneously over time, often arising as a response to convoluted and violent external environments. Community organization, leadership, cooperation, necessity of progress and protection from violence and disasters are the core factors that empower communities to organize and take actions in the built or natural environment. Often these initiatives are created to benefit the neighbourhood and closer communities. Of special attention are the Local Nature Intervention Projects (LNIP) created and sustained by residents of these settlements who aim to recover the natural environment that has been depleted or misused by locals. Although these projects are part of a well-defined network of green infrastructure that is copiously adapted by residents, there has been limited attention to these initiatives in the urban planning and policy research that could help understand how LNIP could be integrated into the city green infrastructure GB. Furthermore, there is a lack of studies that examine the potential of such initiatives and the effects of these community-led interventions on the health and wellbeing of residents.

A longitudinal study has been carried out to understand the underlying action systems in which residents of Medellin communes create appropriation over the natural environment through LNIP and the impact of those projects in the community health and wellbeing. From 2016, data from the residents has been collected in the Medellin communes of Villatina (commune 8) and Eduardo Santo (commune 13). Results from this study have shown important insights of the mechanisms that residents use to cope with the mental stress caused by extreme violent events, disasters and poverty. Two salient factors were found to be at the core of the process of adaptation to extreme events: Nature interaction and community activism.

Nature interaction is enabled through the LNIP, which is part of a green network infrastructure that is often located at the fringes of residual green areas (i.e. nearby roads or areas next to water streams). Although these green areas are randomly located, each of them has a unique function given by the person or family that takes care of the place. In some cases, activities for nature contemplation and community gathering are displayed in the landscape setting, which provide the unique characteristics of the place. In other cases, the green infrastructure is used as productive lands to farm and grow trees. Other infrastructures might reserve spaces for flower gardens. The random appropriation of residual green spaces has been occurring for the last thirty years showing important signs of nature preservation, environmental agency and activism from the communities. However, since these initiatives have emerged from a personal initiative towards nature preservation, they have evolved spontaneously over time rarely receiving external financial support.

Results of this study have shown that the LNIP purpose is twofolded. When communities become conscious of the environment they live in—i.e. extremely violent, and convoluted environments—they search for the most adequate coping alternatives that provide them with a greater degree of control over their lives. This is often manifested in the intervention of the built or natural environment. When the interventions are in the natural environment, the community uses the natural landscape with the purpose of finding financial alternatives provided by farming. These lands are used as a safe place for family and community members where they would gather to experience outdoor nature, and to generate a sense of wellbeingness provided by the connection to the natural environment.

Findings also showed that communities participating in LNIP develop a greater sense of ownership towards the natural landscape, which occurs over time and is reinforced by the interventions created to take care and preserve the environment. This unfolds an additional social and political dynamic in the role of the communities, which is manifested in the forms of stewardship, leadership and community cooperation. In other words, those engaged in nature-related projects become the stewards of the natural environment and the leaders in the neighbourhood, and in some cases the local nature interventions are turned into learning platforms for children and residents interested in urban farming—as had been found in the Values Hill in Villatina (Fig. 1). The sense of altruism enacted throughout the process of creating and keeping LNIP provides unfolding feelings of self-worth and sense of belonging towards the territory, which could substantially help in improving the sense of health and wellbeing of the communities.

Feelings of emotional stability and wellbeing were also found to be strong in people that constantly interact in these spaces. During the lockdown period caused by the COVID-19 pandemic in April 2020, a series of phone interviews were conducted with residents. The residents taking care of the LNIP manifested to have spent most of their time in nature performing gardening or farming activities (Figs. 2 and 3). The responses showed that people engaged in such activities expressed feelings of "time flying" or being "unaware of the critical situation". An opposing situation was found for those that did not engage in outdoor nature activities, expressing feelings of being overwhelmed, anxious and stressed. When the communities are faced with extreme circumstances, the results suggest that LNIP could have positive impacts on their mental health and wellbeing to help them overcome the situation. However, it is not yet clear whether the sense of wellbeing created by the LNIP has any relationship with the bottom-up nature of these initiatives and sense of ownership that is manifested towards the landscape.

Fig. 1 Values Hill local nature intervention project (Villatina)

Fig. 2 Resident performing farming activities during COVID-19 lockdown (Eduardo Santos)

3 Conclusions and Recommendations

This study has presented a summary of the nature community-led participatory approach that has evolved over time in two of the most violent communes in Medellin. As part of a community-based approach that has developed in response to the external

Fig. 3 Isolation strategies in the local nature intervention projects—Eduardo Santos Neighbour-hood

environment, it has helped to ensure that the natural landscape is part of the self-expression and agency of segregated communities. Leadership, stewardship, financial stability and mental wellbeing are some of the benefits generated through these initiatives.

We recommend

- To generate further insights into how LNIP initiatives could be part of the integral urban green infrastructure of cities for future nature conservation and urban health policies.
- The active participation and agency of communities to maintain the places, should be formally acknowledged.
- External top-down led participatory approaches (i.e. PDS) may not be needed. On the contrary, the environmental management manifested in Villatina and Eduardo Santos neighbourhoods could be considered as an alternative to avoid disempowerment, and dependency of the communities to the local government interventions.
- Continuity and support of community-led LNIP initiatives which positively affect urban health, advance community empowerment, environmental awareness and wellbeing across settlements located in extreme urban environments.

Printed in the United States
by Baker & Taylor Publisher Services